FLORA OF TROPICAL EAST AFRICA

LEMNACEAE

F. N. Hepper

Aquatic herbs, floating or submerged and freely drifting, very reduced in structure to a flat or curved thallus or suborbicular, often minute; stems and leaves undifferentiated. Vascular tissue minimal. Roots present or absent, often solitary, suspended in the water, devoid of root hairs, tip covered by a cap. Vegetative reproduction usual, daughter thallus budding from a lateral pocket, often remaining attached to parent; resting buds (turions) sometimes produced in adverse conditions. Flowering erratic; flowers monoecious; developing in a pouch or pit, enclosed by a spathe or spathe absent. Staminate flowers 1–2; anthers 2-thecous. Pistillate flower solitary; style short; stigma concave; ovules 1–6. Seeds ellipsoid, usually ribbed; endosperm scanty or none.

A small family (duckweeds) of 6 genera, well known in still water throughout the world, with most species in the warmer countries.

Seldom collected in flower; often overlooked, especially the submerged species; they should be preserved in spirit to aid identification and our understanding of this interesting family which has affinity with *Araceae*. Hegelmaier's excellent study (Lemnac. Monogr., Leipzig 1868) laid the foundation for much subsequent work. Daubs (Monogr. Lemnac., Illinois 1965) unfortunately consulted only American herbaria and his account is of limited value. The taxonomy of the family has been surveyed by den Hartog and van der Plas (in Blumea 18: 355–368 (1970)), while interest elsewhere centres on the physiology and chemotaxonomy (cf. McClure & Alston in Am. Journ. Bot. 53: 849–860 (1966)).

Roots present:
- Root solitary on each thallus; thallus floating or submerged . . . . 1. **Lemna**
- Roots numerous; thallus floating . . . . . 2. **Spirodela**

Roots absent:
- Thallus flat; budding pouch triangular, dorsiventrally flattened, opening by a transverse split:
  - Thallus submerged; inflorescences 2 . . . . 3. **Wolffiopsis**
  - Thallus floating, only stipe submerged; inflorescence 1 . . . . 4. **Pseudowolffia**
- Thallus ± globular; budding pouch funnel-shaped, with a circular opening . . . . . . 5. **Wolffia**

1. LEMNA

L., Sp. Pl.: 970 (1753) & Gen. Pl., ed. 5: 417 (1754); Hegelm., Lemnac. Monogr.: 134 (1868); N.E. Br. in F.T.A. 8: 201 (1901); Daubs, Monogr. Lemnac.: 16 (1965)

Aquatics, submerged or more usually floating in great numbers; individual thalli cohering. Thalli oblong to ovate, ± asymmetrical, thin or spongy. Daughter thallus budding from lateral pocket. Root solitary, devoid of hairs, shortly sheathed near point of origin; sheath simple or laterally winged; apex of root sheathed by a conspicuous root-cap, tip acute or obtuse. Flowering erratic; floral pocket lateral with spathe enclosing 2 staminate and 1 pistillate flowers. Fruit (1–6)-seeded. Seed usually ribbed.

About 9 species, cosmopolitan.

Thalli submerged (except for occasional free-floating fertile thalli); thalli forming branching chains, each thallus stalked, margins toothed . . . . . 1. *L. trisulca*
Thalli free-floating on water-surface:
Root-caps obtuse; sheath of root at junction with thallus not winged:
Thallus ± hemispherical with large inflated cells beneath . . . . 2. *L. gibba*
Thallus ± flat:
Thallus with large cells beneath . . . . . 2. *L. gibba*
Thallus with small cells and small airspaces; root arising in a groove . . . . . . 3. *L. minor*
Root-caps acute; sheath of root at junction with thallus winged; thallus flat, thin . . . . 4. *L. perpusilla*

1. **L. trisulca** *L.*, Sp. Pl.: 970 (1753); Hegelm., Lemnac. Monogr.: 134, t. 5, 6, figs. 1–18 (1868); E. & P. Pf. 2, 3: 164, figs. 104, 106/A, B (1889); den Hartog in Gorteria 2: 68, fig. 1 (1964); Daubs, Monogr. Lemnac.: 32, t. 6 (1965). Types: Europe, specimens in Hort. Cliff. (BM, syn.!) & Herb. Linn. (LINN, syn.!)

Submerged aquatic. Thalli thin, oblong-lanceolate, 5–10 mm. long, with a stipe almost as long again, 2–4 mm. wide, margin ± toothed in the upper third, entire and undulate below. Daughter thallus budding in pouch situated in lower third, each remaining attached by stipe, thus forming a lattice. Root, if present, solitary; cap acute. Fertile thalli few, non-stipitate, much smaller than non-fertile thalli, free-floating at water-surface; flowering pouch produced near base of thallus. Spathe present, enclosing 2 ♂ and 1 ♀ flower. Staminate flowers protandrous, each with one curved stamen at different stages of development, making direct contact with style of another plant. Pistillate flower with 1 ovule. Seed oblong, ribbed. Fig. 1/1.

UGANDA. Kigezi District: Lake Mutanda, Nov. 1940, *Eggeling* 4215! & Kumba, 1 July 1969, *Lye* 3441! & Lake Bunyonyi, Dec. 1938, *Chandler & Hancock* 2600!
KENYA. Lake Naivasha, July 1963, *Suttie* H241/62/2!; Machakos District: Makindu, 1 July 1954, *Rayner* 559!; Masai District: Ol Tukai, 9 Feb. 1964, *Verdcourt* 3975!
TANGANYIKA. Arusha District: Lake Duluti [Balbal], 14 Nov. 1901, *Uhlig* 494!; Lushoto District: Pangani R., 13 Sept. 1951, *Greenway* 8715d!
DISTR. **U**2; **K**3–6; **T**2, 3; not recorded elsewhere in tropical Africa, mainly in northern temperate zones
HAB. Freshwater lakes and pools; 500–1900 m.

2. **L. gibba** *L.*, Sp. Pl.: 970 (1753); Hegelm., Lemnac. Monogr.: 145, t. 11–13 (1868); E. & P. Pf. 2, 3: 164, fig. 101/C, 106/D (1889); N.E. Br. in F.T.A. 8: 201 (1901); Daubs, Monogr. Lemnac.: 17, t. 8 (1965). Type: Europe, unspecified

Thallus free-floating, surface broadly ovate, yellow-green, asymmetrical, slightly convex, 3–5(–6) mm. long, 2–4(–5) mm. wide, almost hemispherical below and spongy with large inflated hyaline cells and airspaces (even in flat form with large cells evident). Daughter thalli budding laterally, 2 or 3

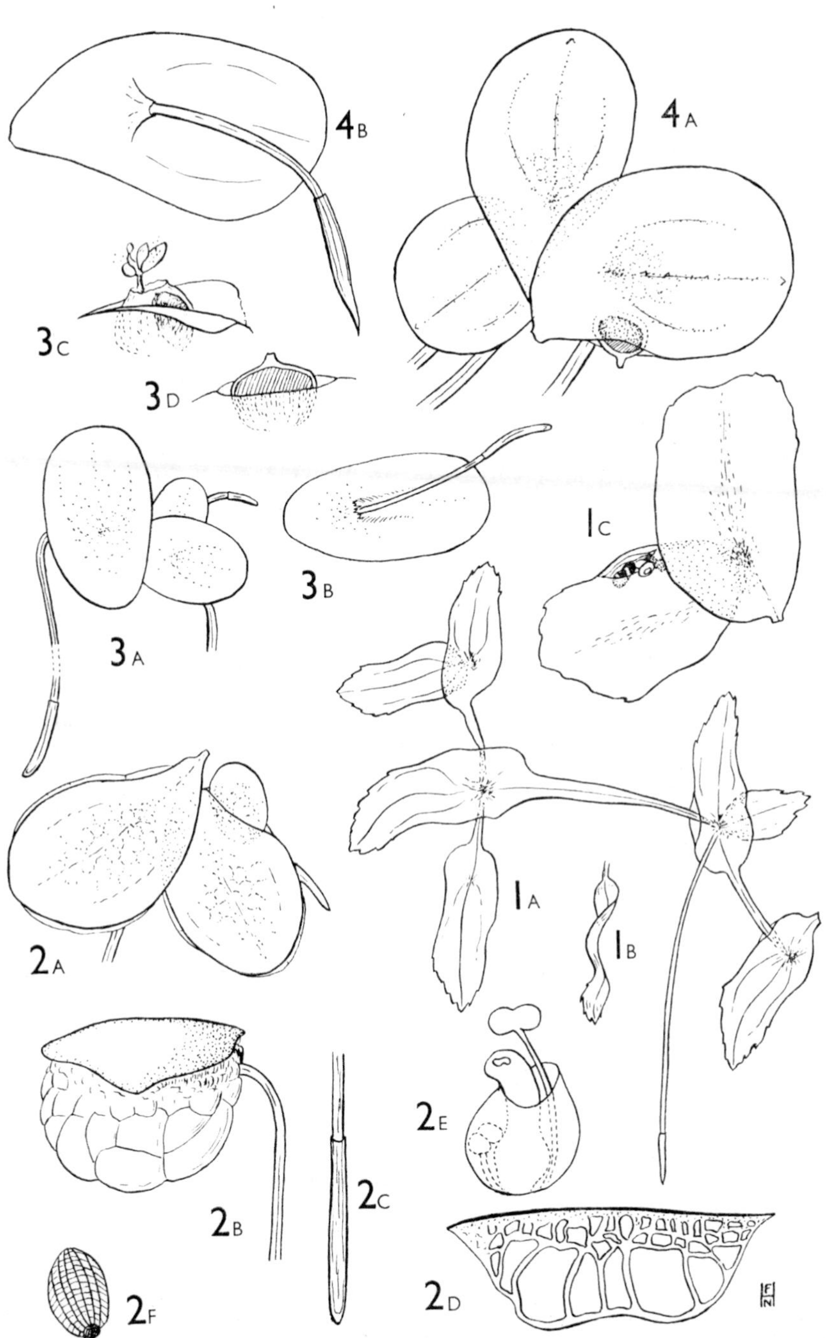

Fig. 1. *LEMNA TRISULCA*—**1A,** sterile thalli, × 4; **1B,** twisted thallus, × 4; **1C,** fertile thalli, × 18. *L. GIBBA*—**2A,** sterile thalli, × 8; **2B,** lateral view, × 10; **2C,** root cap, × 10; **2D,** transverse section, of thallus × 18; **2E,** flower, × approx. 27; **2F,** seed, × approx. 12. *L. MINOR*—**3A,** sterile thalli, × 8; **3B,** ventral surface of thallus, × 8; **3C,** dehisced anthers, × approx. 27; **3D,** fruit, × approx. 27. *L. PERPUSILLA*—**4A,** thalli, one with fruit, × 8; **4B,** ventral surface, × 8. 1A, B, from *Richards* 20226; 2A, D, from *Richards* 23320; 2B, C, from *Richards* 20167; 2E, F, after Daubs (1965); 3A, from *A. S. Thomas* 2410; 3B, C, D, from *Melville* s.n.; 4A, B, from *Richards* 21000.

remaining attached with very short stipes. Root solitary; sheath unwinged; root-cap 5 mm. long, tip obtuse (rarely acute). Floral cavity lateral. Spathe enclosing 2 1-staminate flowers and 1 pistillate. Fruits narrowly winged above, 1–2-seeded. Seed ribbed and transversely striate. Fig. 1/2, p. 3.

KENYA. Machakos District: Kibwezi, Chai Springs, *J. Brown*!; Masai District: Namanga River Hotel, 13 Jan. 1954, *Greenway* 8795! & Ondoni R., Magadi Junction, 1926, *Prescott Decie*!

TANGANYIKA. Maswa District: Moru Kopjes, 9 Feb. 1968, *Greenway, Kanuri & Braun* 13158!; Arusha District: E. of Mt. Meru, Ngongongare, 10 Dec. 1968, *Richards* 23320!; Singida Lake, 27 Apr. 1962, *Polhill & Paulo* 2198!

DISTR. **K**4–7; **T**1, 2, 5; North Africa, Ethiopia, South Africa and widespread through most of the world

HAB. Pools and still water, covering the surface; 800–1540 m.

VARIATION. The typical spongy form is very easily identifiable, but the flat form is difficult to separate from *L. minor* (see note under that species).

3. **L. minor** *L.*, Sp. Pl.: 970 (1753); Hegelm., Lemnac. Monogr.: 142, t. 9–10 (1868); E. & P. Pf. 2, 3: 164, 154, fig. 101/B, 107/A, B, 108–9/A (1889); P.O.A. C: 133 (1895); N.E. Br. in F.T.A. 8: 202 (1901); Daubs, Monogr. Lemnac.: 21, t. 7 (1965). Type: Europe (LINN, ? holo.!)

Free-floating aquatic. Thallus dark green, symmetrical, flat, oblong-ovate, 2·5–6 mm. long, 1·5–4 mm. wide, rather obscurely 3-nerved, without large airspaces evident beneath. Root solitary, arising along a shallow groove; sheath unwinged; root-cap obtuse. Floral pocket lateral; spathe open at the top. Staminate flowers 2, each with a single stamen. Pistillate flower solitary. Seed solitary, not ribbed, reticulate. Fig. 1/3, p. 3.

UGANDA. Kigezi District: Lake Mutanda, 21 Aug. 1938, *A. S. Thomas* 2410! & Rubanda county, Muko school, 11 Dec. 1968, *Lye & Lester* 1019!

KENYA. Kiambu/Machakos District: Fourteen Falls, 6 Sept. 1952, *Verdcourt* 718b!

DISTR. **U**2; **K**4; Ethiopia, Congo, South Africa and widespread elsewhere

HAB. Surface of pools at cooler altitudes; up to 1800 m.

VARIATION. In the dried state *L. minor* is difficult to distinguish from the flat form of *L. gibba*. The latter has more obvious airspaces, which can be seen beneath and the Kenya specimens cited above may in fact be referable to flat, *L. gibba*. Field workers should make careful comparisons between them.

4. **L. perpusilla** *Torrey*, Fl. New York 2: 245 (1843); Hegelm., Lemnac. Monogr.: 139, t. 6/19, 20, t. 7/18, 19 (1868); F.P.S. 3: 284 (1956); Daubs, Monogr. Lemnac.: 25, t. 9 (1965). Type: U.S.A., New York, Staten I., *J. Torrey* (BM, MO, iso.!)

Thallus free-floating, pale green, ovate, variable in size, 1·5–4 mm. long, 0·7–2·5 mm. wide, asymmetrical, upper surface with or without an apical papilla and 1 or more along the midline, obscurely 3-nerved. Daughter thallus very shortly stipitate, several individuals adhering closely. Root solitary; sheath winged below the thallus; root-cap acute. Flowering cavity lateral; spathe open. Staminate flowers 2; anthers 2-thecous. Pistillate flower 1; style concave. Fruit ribbed, 1-seeded. Fig. 1/4, p. 3.

UGANDA. Toro District: Kasenyi, 16 Dec. 1967, *Lock* 67/172!; Mengo District: Kakinzi, 14 Oct. 1969, *Lye* 4413! & Singo, Bukomero, Sept. 1932, *Eggeling* 558!

KENYA. Masai District: Amboseli, Ol Tukai, 14 May 1961, *Verdcourt* 3117!; Kwale District: between Samburu and Mackinnon Road, Taru, 3 Sept. 1953, *Drummond & Hemsley* 4141!; Kilifi District: between Mariakebuni and Marafa, 23 Nov. 1961, *Polhill & Paulo* 845!

TANGANYIKA. Musoma District: Grumeti R., Kirawira, 1 Sept. 1962, *Greenway* 10796!; Lushoto/Tanga Districts: Ngua, 19 July 1950, *Verdcourt* 290!; Iringa District: Ruaha National Park, 15 Jan. 1966, *Richards* 21000!

DISTR. **U**2, 4; **K**5–7; **T**1–4, 6, 7; widespread in Africa and throughout the tropical and warmer parts of the world

HAB. Covering surface of pools in a continuous sheet or fewer individuals in backwaters, rock pools and rice fields; 0–1600 m.

SYN. *L. aequinoctialis* Welw., Apont. Phytogeo.: 578 (1859); P.O.A. C: 422 (1895); N.E. Br. in F.T.A. 8: 203 (1901); F.P.S. 3: 284 (1956); Giardelli in Darwinia 11: 584–590 (1960). Type: Angola, *Welwitsch* 206 (BM, K, iso.!)

L. paucicostata Engelm. in Gray, Man. Bot., ed. 5: 681 (1867); Hegelm., Lemnac. Monogr.: 139, t. 8 (1868); P.O.A. C: 422 (1895); N.E. Br. in F.T.A. 8: 202 (1901); Jumelle in Fl. Madag. 32: 2 (1937); Maheshwari & Kapil in Am. Journ. Bot. 50: 677, 907 (1963); McClure in Am. Journ. Bot. 59: 849 (1966); Hepper in F.W.T.A., ed. 2, 3: 129 (1968). Type: not specified

L. perpusilla Torrey var. *trinervis* Austin in Gray, Man. Bot., ed. 5: 479 (1867). Type: U.S.A., Pennsylvania, Wayne County, *Austin* (MO, iso.)

L. angolensis Hegelm., Lemnac. Monogr.: 141 (1868). Type: Angola, *Welwitsch* 206 (BM, K, iso.!)

L. paucicostata Engelm. var. *membranacea* Hegelm. in Lemnac. Monogr.: 141 (1868)

L. trinervis (Austin) Small, Fl. SE. United States: 230 (1903); Daubs, Monogr. Lemnac.: 27, t. 13 (1965)

VARIATION. In Africa the very small form has been known as *L. aequinoctialis* and the larger one as *L. paucicostata*. However, the type of *L. perpusilla* is very small too, and there appears to be no characters to distinguish them as separate species. The larger, very membranous individuals with 3 rather distinct nerves are regarded by some as *L. trinervis*.

2. SPIRODELA

Schleid. in Linnaea 13: 391 (1839); Hegelm., Lemnac. Monogr.: 147 (1868); Daubs, Monogr. Lemnac.: 8 (1965)

Floating aquatics. Thallus discoid to oblong, 3–10 mm. long, 1·2–8 mm. wide. Daughter thallus budding from lateral slit, remaining attached to parent by slender stipe, several individuals often connected. Roots 2–numerous, usually from swollen portion beneath thallus; root-cap acute. Turions (resting buds) produced in adverse conditions. Flowers rarely produced, unisexual; spathe enclosing 2–3 ♂ and 1 ♀ flowers in lateral slit pouch; ♂ flowers each with 1 stamen; pistillate flower bearing 1–4 ovules.

Cosmopolitan genus with 5 species, easily recognised by the presence of several to numerous roots.

S. polyrhiza (*L.*) *Schleid.* in Linnaea 13: 392 (1839); Hegelm., Lemnac. Monogr.: 73, 151, t. 13/10–16, t. 14, 15 (1868); P.O.A. C: 422 (1895); Jumelle in Fl. Madag. 32: 3 (1937); Lawalrée in B.S.B.B. 77: 27 (1945); Troupin in Expl. Parc Nat. Garamba 1: 154 (1956); Maheshwari in Nature 181: 1745 (1958); Daubs, Monogr. Lemnac.: 10, t. 3 (1965); Hepper in F.W.T.A., ed. 2, 3: 129, fig. 361/B (1968). Type: Europe (not in LINN).

Free-floating aquatic. Thallus orbicular-ovate, 3–8 mm. long, almost as broad, with 5–11 conspicuous nerves, sometimes tinged pink above, usually purplish beneath. Daughter thallus budding from near point of root-insertion in a slit in the mother thallus. Roots 5–15, arising from greatly thickened basal part of thallus; root-cap acute. Rootless resting bud (turion) produced in adverse conditions, 2 mm. in diameter. Flowers seldom produced, surrounded by a small open spathe in a lateral slit-pouch, with 1 pistillate flower, and 2–3 staminate flowers each consisting of a single stamen. Fruit slightly wing-margined. Seeds 1–2. Fig. 2/1.

UGANDA. Toro District: Nsonge R. north of Lake George, 28 Oct. 1967, *Lock* 67/115!; Busoga District: near Jinja, Kyrinye, July 1953, *Lind* 178!; Mengo District: Namanve Swamp, Sept. 1932, *Eggeling* 559!

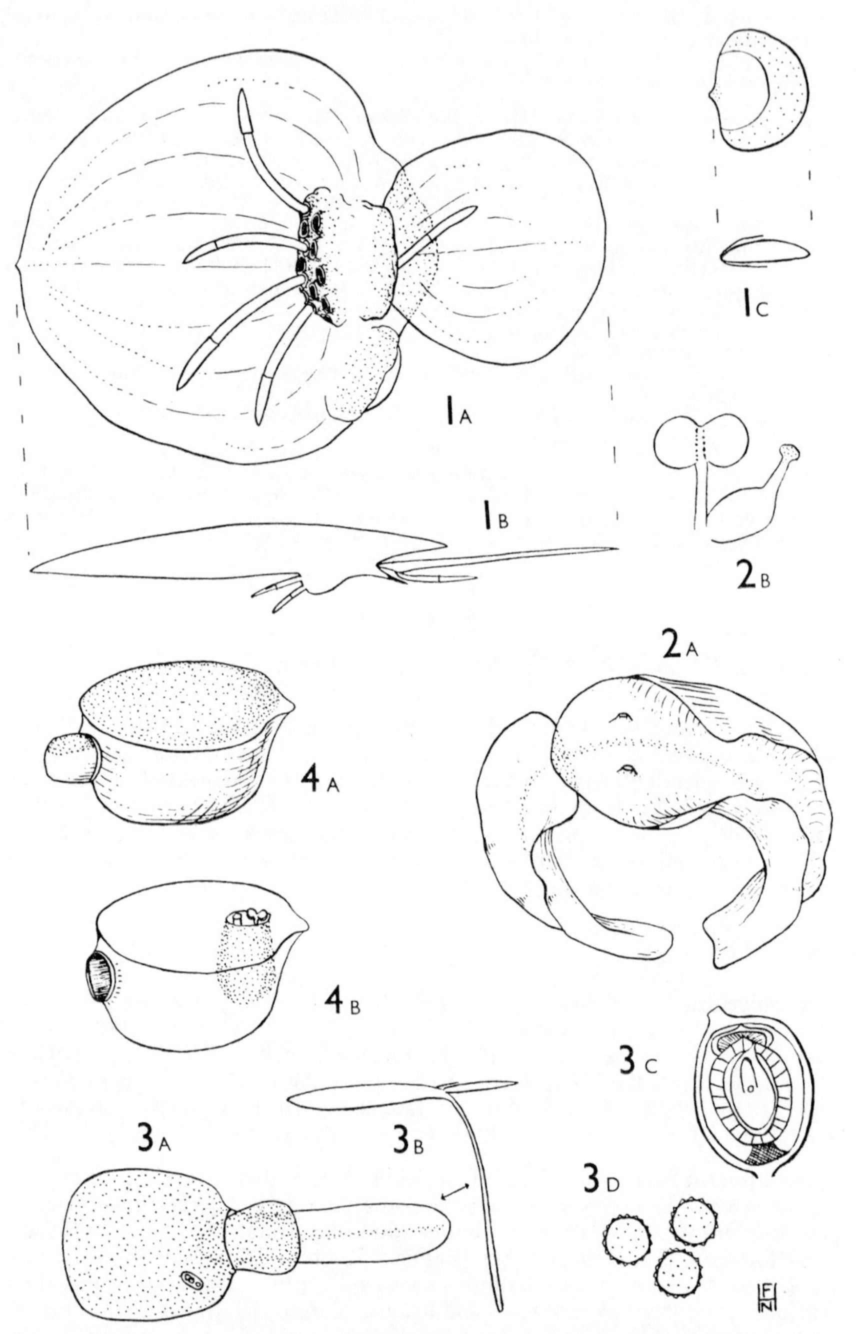

Fig. 2. *SPIRODELA POLYRHIZA*—**1A,** ventral surface of sterile thalli, × 8; **1B,** cross-section; **1C,** resting bud, × 6. *WOLFFIOPSIS WELWITSCHII*—**2A,** fertile thallus, × 12; **2B,** stamen and pistil, × 20. *PSEUDOWOLFFIA HYALINA*—**3A,** dorsal view of fertile thallus, × 12; **3B,** thallus in floating position, × 12; **3C,** median section of fruit and seed, × approx. 25; **3D,** pollen grains typical of all *Lemnaceae*, × approx. 160. *WOLFFIA ARRHIZA*—**4A,** sterile thallus with bud, × 20; **4B,** fertile thallus, × 20. 1A, B, C, from *Brenan* 4995; 2A, from *Hall* 3036; 2B, after Daubs (1965); 3A, B, from *Chevalier* 1143; 3C, D, after Daubs (1965). 4A, B, after *Hall* 3037.

KENYA. Masai District: Amboseli National Park, 4 Aug. 1956, *Milne-Redhead & Taylor* 7048!; Tana R. District: R. Tana near Wema, 17 Feb. 1972, *Gillett & Kibuwa* 19956!

TANGANYIKA. Bukoba District: Lake Burigi [Urigi See], Chitara, 24 Mar. 1913, *Braun* in *Herb. Amani* 5553!; Mbulu District: Lake Manyara, NW. side, 10 Mar. 1964, *Greenway & Kanuri* 11342!; Lushoto District: Pangani R., Maurui, 13 Sept. 1951, *Greenway* 8715!

DISTR. **U**2–4; **K**6, 7; **T**1–3; cosmopolitan

HAB. Still water, numerous individuals often cohering or floating together on the surface of the water; 450–1200 m.

SYN. *Lemna polyrhiza* L., Sp. Pl.: 970 (1753); P.O.A. C: 422 (1895); N.E. Br. in F.T.A. 8: 201 (1901); F.P.S. 3: 283 (1956)

3. **WOLFFIOPSIS**

den Hartog & van der Plas in Blumea 18: 366 (1970)

Small aquatic herb, submerged except for a portion of each thallus at water-level where a few stomata are present. Thallus thin, undulate and falcate; daughter thallus budding from a flattened-conical pouch, sometimes forming a chain of several individuals in still water. Roots absent. Flowers rarely produced in 2 pits at either side of mid-line on upper surface, 1 ♂ flower (with 1 stamen) and 1 ♀ flower occurring in each pit. Style short; ovule 1, orthotropous. Fruit with persistent style. Seed spherical-turbinate; testa reticulate.

A monotypic genus occurring in Africa and America in tropical latitudes.

W. welwitschii (*Hegelm.*) *den Hartog & van der Plas* in Blumea 18: 366 (1970). Type: Angola, near Quizembo, *Welwitsch* 209 (BM!, K!, MO, iso.)

Floating thallus usually curved in a semicircle, the broadest part at the surface of the water, thin, concave, the inturned margins slightly undulate, 2–5 mm. long, ± 2(–4) mm. broad. Daughter thallus budding from the broader end and curving beneath the older thallus thereby ± completing a circle. A pair of flowering pouches developed, one on each side of the mid-line at the water-surface, with 1 pistillate flower with short style, and 1 staminate flower with 1 stamen with very short thick filament; flowers not enclosed in a spathe. Seed almost spherical to turbinate, ± 0·8 mm. long, reticulate. Fig. 2/2.

UGANDA. Kigezi District: Rubanda, Muko, 11 Dec. 1968, *Lye* 1021!

TANGANYIKA. Lushoto District: Maurui, 13 Sept. 1951, *Greenway* 8715c!

DISTR. **U**2; **T**3; Sudan west to Senegal and Angola; also reported from Venezuela, Cuba, S. Domingo, Guatemala and Ecuador

HAB. Pools, apparently in and near the forest zone; 450–2100 m.

SYN. *Wolffia welwitschii* Hegelm. in J.B. 3: 114 (1865) & Lemnac. Monogr.: 130, t. 4/1–10 (1868); E. & P. Pf. 2, 3: 164, fig. 105/c (1889); N.E. Br. in F.T.A. 8: 205 (1901); F.P.S. 3: 284 (1956)
Wolffiella welwitschii (Hegelm.) Monod in Mém. Soc. Hist. Nat. Afr. Nord, Hors-Sér. 2: 229, fig. 1–41 (1949); Daubs, Monogr. Lemnac.: 35, t. 16 (1965); Obermeyer-Mauve in S. Afr. Journ. Sci. 62: 277–278, fig. 1 (1966); Hepper in F.W.T.A., ed. 2, 3: 127, fig. 361/D (1968)

4. **PSEUDOWOLFFIA**

den Hartog & van der Plas in Blumea 18: 365 (1970)

Small aquatic herbs, floating on the surface of pools. Thallus flat, with air-spaces, upper surface provided with numerous stomata. Daughter thallus budding from a symmetric pouch opening by a transverse slit with a long membranous, ± widened median stipe on the underside. Flowers in a solitary lateral or almost median pit.

A genus with 3 species, all occurring in Africa.

P. hyalina (*Del.*) *den Hartog & van der Plas* in Blumea 18: 366 (1970). Type: Egypt, Damiata, collection not indicated (? MPU, holo.)

Floating aquatic with thallus in 2 parts: a nearly opaque green floating portion, broadly elliptic, up to 2 mm. long and 1·5 mm. broad, and a hyaline oblong appendage 2–4(–6) mm. long, 1(–1·5) mm. broad, at right-angles to the other and suspended in the water. Daughter thallus developing in slit-pouch between the 2 parts and arising from the thicker portion. Flowers rarely produced, flower-pit solitary near pouch, to one side of mid-line, bearing 1 pistil and 1 stamen. Fig. 2/3, p. 6.

UGANDA. Toro District: Queen Elizabeth National Park, Mweya Peninsula, 19 Mar. 1968, *Lock* 68/46!
KENYA. Masai District: Amboseli, Ol Tukai, 14 May 1961, *Verdcourt* 3116!
TANGANYIKA. Masai District: Ngorongoro Crater, Lake Magadi, 5 July 1966, *Greenway & Kanuri* 12538!; Arusha District: Kinandia Lake, 23 Mar. 1966, *Greenway & Kanuri* 12483!; Handeri District: Segera, 8 Mar. 1956, *Greenway* 8974b!
DISTR. **U**2; **K**6; **T**2, 3, 5; Egypt, Sudan and Central African Republics, N. Cameroun, N. Nigeria and Mali
HAB. In pools and slowly running water in drier regions of Africa; 300–1700 m.

SYN. *Lemna hyalina* Del., Fl. Aegypt.: 75 (1813)
Wolffia delilei Schleid. in Linnaea 13: 390 (1839); N.E. Br. in F.T.A. 8: 204 (1901), *nom. superfl.* Type: as for species.
W. hyalina (Del.) Hegelm., Lemnac. Monogr.: 128, t. 4, figs. 11–19 (1868); E. & P. Pf. 2, 3: 164, fig. 101/F, G (1889); P.O.A. C: 423 (1895); F.P.S. 3: 284 (1956)
Wolffiella hyalina (Del.) Monod in Mém. Soc. Hist. Nat. Afr. Nord, Hors-Sér. 2: 242 (1949); Hepper in F.W.T.A., ed. 2, 3: 127 (1968); Jovet-Ast in Bull. I.F.A.N., sér. A, 30: 834, fig. 2 (1968)

5. WOLFFIA

Schleid. in Linnaea 13: 389 (1839); Hegelm., Lemnac. Monogr.: 121 (1868); N.E. Br. in F.T.A. 8: 203 (1901); Daubs, Monogr. Lemnac.: 41 (1965)

Minute free-floating aquatics. Thallus almost ellipsoid, ± 1 mm. in diameter, flattened on the upper surface at water-level. Daughter thallus produced in a circular lateral pit, soon separating from parent. Roots absent. Flowers in a pit on the upper surface of the median line, 1 staminate, 1 pistillate; spathe absent.

Cosmopolitan genus with 7 species, lacking roots or appendages.

W. arrhiza (*L.*) *Wimmer* in Fl. Schles., ed. 3: 140 (1857); Hegelm., Lemnac. Monogr.: 124, t. 2, figs. 6–17, t. 3/1–12 (1868); P.O.A. C: 422 (1895); E. & P. Pf. 2, 3: 164, figs. 101/E, D, 103, 105/A, B, 106/E, F (1889); Daubs, Monogr. Lemnac.: 48, t. 18 (1965); Hepper in F.W.T.A., ed. 2, 3: 129 (1968). Types: Italy & France, *Du Chesne*

Minute aquatic 1–1·5 mm. long, rather less in breadth and depth, upper surface flattened and dark green, paler elsewhere. Flowers opening on to surface from single median pit. Stamen solitary, with filament 0·5 mm. long. Style short; stigma concave. Fruit ellipsoid, 0·7 mm. long, erect. Fig. 2/4, p. 6.

Uganda. Ankole District: 104 km. Masaka–Mbarara, 6 Dec. 1967, *Lock* 67/156!; Kigezi District: Muko, 11 Dec. 1968, *Lye & Lester* 1020!
Kenya. Kisumu, 19 Dec. 1958, *McMahon* 215!; Masai District: Amboseli, Ol Tukai, 14 May 1961, *Verdcourt* 3107!; Kwale District: Taru, 5 Sept. 1953, *Drummond & Hemsley* 4164!
Tanganyika. Pare District: R. Pangani (Ruvu), 5 Nov. 1955, *Milne-Redhead & Taylor* 7237!; Lushoto District: Amani, 24 Sept. 1936, *Greenway* 4624!; Uzaramo District: Dar es Salaam, Feb. 1874, *Hildebrandt* 1236!
Distr. **U**2; **K**5–7; **T**3, 6; scattered throughout Africa and probably overlooked; widespread in the Old World
Hab. Ditches and pools, sometimes occurring abundantly enough to form a scum; 0–2100 m.

Syn. *Lemna arrhiza* L., Mant. Pl. Alt.: 294 (1771)
Wolffia michelii Schleid., Beitr. Bot.: 233 (1844); N.E. Br. in F.T.A. 8: 205 (1901), *nom. superfl.* Type: as for species

INDEX TO LEMNACEAE